41 Activities in Basic Money Management

SECOND EDITION

Nancy Lobb

J. WESTON
WALCH
PUBLISHER

Portland, Maine

User's Guide
to
Walch Reproducible Books

As part of our general effort to provide educational materials that are as practical and economical as possible, we have designated this publication a "reproducible book." The designation means that purchase of the book includes purchase of the right to limited reproduction of all pages on which this symbol appears:

Here is the basic Walch policy: We grant to individual purchasers of this book the right to make sufficient copies of reproducible pages for use by all students of a single teacher. This permission is limited to a single teacher and does not apply to entire schools or school systems, so institutions purchasing the book should pass the permission on to a single teacher. Copying of the book or its parts for resale is prohibited.

Any questions regarding this policy or requests to purchase further reproduction rights should be addressed to:

Permissions Editor
J. Weston Walch, Publisher
321 Valley Street • P.O. Box 658
Portland, Maine 04104-0658

1   2   3   4   5   6   7   8   9   10
ISBN 0-8251-3945-7

Contents

Unit 4: Automating Your Money (Electronic Banking)

Unit 5: Managing Your Money (Credit and Loans)

To the Teacher

41 Activities in Basic Money Management is a reproducible resource book designed to help you teach your students the basic concepts needed to handle money wisely. The student material is written on a third-grade reading level, with a third-grade level of math skills required. The interest level of the material is age 12 to adult. Thus, it would be appropriate for special education students in grades 6–12, for Adult Basic Education classes, or for English-as-a-Second-Language programs.

The concepts presented in *41 Activities in Basic Money Management* are divided into five units. These are:

> Unit 1: Earning Your Money (The Paycheck)
> Unit 2: Spending Your Money (A Checking Account)
> Unit 3: Saving Your Money (A Savings Account)
> Unit 4: Automating Your Money (Electronic Banking)
> Unit 5: Managing Your Money (Credit and Loans)

The first activity in each unit presents four to six vocabulary words basic to the understanding of that unit. The following activities provide a simple, step-by-step teaching sequence of the unit concepts. The last activity in each unit is a review or quiz, which assesses mastery of each key concept taught in that unit.

A comprehensive selection of additional instructional activities for each unit of study is provided in the following section. For each unit, a brief listing of Internet search terms is also included. Students can enter these terms into an Internet search engine to find links with more information on a given topic. To find more information on the Internet about money management topics in general, enter the following search terms into a search engine: *personal finance, financial planning, banking and finance, credit unions, federal deposit insurance.* An answer key is also provided and can be found at the back of this book.

41 Activities in Basic Money Management contains both a teacher section and a student section. Reproducible pages are identified by the copyright line with a flame logo at the bottom. They may be copied and distributed to students. In addition, activities for which you may choose to allow students to use calculators are marked with this symbol:　　　　on pages *vii–xvi.*

You will find that *41 Activities in Basic Money Management* is easy to use. Students will enjoy the material as well, as it clearly relates to a subject that is very important to them. It is certainly critical that all of your learners learn how to handle their money wisely at a young age, so that they can make good decisions as they begin working and living on their own.

NOTE: Throughout the text, references are made to using "bank" services. The term *bank* could also include credit unions. Explain to students the difference between a bank and a credit union. (Joining a credit union can be advantageous—especially in terms of lower fees and interest rates—if one is eligible.)

Additional Instructional Activities

Unit 1: Earning Your Money (The Paycheck)

To find more information on the Internet about paychecks, enter the following search terms into a search engine: *paycheck, paycheck deductions, paycheck calculator, estimated paycheck.*

Activity 1: Paycheck Vocabulary

1. Give a spelling test on the four vocabulary words and phrases.
2. Have students write an original paragraph using all four terms.
3. For more advanced students, you may wish to include some of these terms: *payroll savings plan, Social Security, automatic deposit, annuity, overtime pay, insurance, disability, contributions, minimum wage, union.*

Activity 2: Finding Number of Hours Worked

1. Construct similar exercises if students need more practice.
2. On the board, write various hours worked for several people. Have students determine who worked the longest or the shortest hours. For example: Dentist—9 A.M. to 5 P.M.; Nurse—3 P.M. to 11 P.M.; Professor—8 A.M. to 6 P.M., etc.

Activity 3: Finding Gross Pay

1. Have students investigate hourly rates of pay at several local businesses that hire young people for after-school work (for example, fast food restaurants, car washes, grocery stores, etc.). Make up problems to determine how much students can earn working 10, 15, or 20 hours a week at these businesses. Or, have students determine how much you could earn working 40 hours per week during summer vacation.
2. Determine the hourly rate of pay for jobs in which your students are interested as possible career choices. Then, have students figure how much they would earn per week, per month, and per year working at these jobs.

Activity 4: Adding Deductions

There are two classes of deductions. Federal withholding tax, state income tax, and FICA (or Social Security) are withheld from earnings and are mandatory. Other deductions may be optional, such as health insurance, dental insurance, life insurance, union dues, parking fees, annuities, disability insurance, savings, Christmas clubs, charitable contributions (e.g., United Way), uniforms, and so on.

1. Discuss why federal withholding tax, state income tax, and FICA are mandatory deductions.

2. Have students ask working people they know what kinds of deductions (not amounts) they have taken out of their paychecks. Make a list of these deductions as a class.

3. Discuss the pros and cons of a payroll savings-deduction plan.

4. Discuss the withholding form you must sign for tax purposes when starting a new job. Obtain a copy of this form and discuss how to fill it out.

Activity 5: Finding Net Pay

1. Obtain additional paycheck forms for students to practice reading.

2. Help students determine the percentage of gross income that is taken by deductions, using the examples on this exercise.

Activity 6: Paycheck Review

Activity 6 can be used as a review or quiz. It includes questions on each topic covered in Activities 1–5.

Unit 2: Spending Your Money (A Checking Account)

To find more information on the Internet about checking accounts, enter the following search terms into a search engine: *banking associations, banks, bank rates, credit units, telephone banking.*

Activity 7: Checking Account Vocabulary

1. Give a spelling test on the six vocabulary words and phrases.

2. Have students write an original paragraph using all six terms.

3. For more advanced students, you may wish to include some of the following terms: *joint account, individual account, overdraft, service charges, stop-payment order, endorsement, insufficient funds, a "bounced" check.*

Activity 8: Opening a Checking Account

1. Discuss the advantages of having a checking account:
   (a) It's not necessary to carry a lot of cash when shopping.
   (b) A check can be cashed only by the person to whom it is written.
   (c) A canceled check is proof of payment.
   (d) It is easier to keep track of how you spend your money.
   (e) It is safer to send checks through the mail than to send cash.
   (f) Your money is safe in a checking account.

2. Ask a bank or credit union officer to visit your class to explain services offered by his or her institution.

3. Discuss the differences between various checking accounts. Some charge a small fee (often about $.15) for each check you write. Some charge a monthly service charge (often about $7). Others are free with a minimum balance (which might be about

$1,000) in your checking account; if your balance goes below this amount, you will pay a service charge. Most banks offer free pamphlets explaining the different checking accounts they offer. Obtain one of these. Have students discuss the information and decide which type of account would work best for them at this time.

4. Have students figure out what it would cost to use a checking account that charges a fee for each check written. For example, a fee of $.15 per check means $1.50 to write 10 checks.

5. Discuss the different types of checks available. Checks can come with stubs, with "carbon" copies, or with a checkbook register. Talk about pros and cons of each style of check.

6. Explain to students that checks may be plain or decorated with a variety of scenes or designs. These "designer" checks are usually more expensive. Students need to ask about the charge for any type of check they order.

Activity 9: Writing Checks

Activity 10: Writing More Checks

1. Discuss what happens to a check after it is given to a clerk in a store.

2. Discuss the meaning of the numbers found on a check.

    (a) The check number is found in the upper-right-hand corner of the check. This number should be recorded in the checkbook register along with the amount of the check. (See Activity 14.)

    (b) Found immediately below the check number is a number that looks like a fraction (e.g., 9-98/790). The top half is the bank number. Some people use the code "9-98" when listing deposits on a deposit slip. The bottom number (790) is for bank use.

    (c) The numbers along the bottom of the check are numbers used by the bank to process the checks. The last set of these numbers is your account number.

3. Explain to the class how a check can be used to get money for yourself out of an account. To do this, you make a check out to "cash." You will need to go to your own bank or a branch of that bank. The teller will cash the check for you. Caution students that it is best to wait until getting to the bank before writing a check to "cash." If you were to lose the filled-out check, another person might try to cash it.

4. Discuss what should be done if an error is made while writing a check. If you write the wrong date, simply cross out the error and correct it. Then write your initials beside the corrected information.

    If you make a mistake on the amount, you should void the check. To do this, write VOID in large letters across the check. Tear up the check or keep it for your records. Then write a new check.

5. Discuss why it would be best not to make corrections on the amount of a check. (The bank may not accept the check; they may suspect it has been tampered with.)

Activity 11: Making Deposits to a Checking Account

1. Discuss the purpose of deposit slips and what should be done with them. (They should be kept as proof of the deposit being made until they can be checked against the printed bank statement, which may be received quarterly or monthly.)

2. You may be able to obtain additional deposit slips from a local bank to provide practice. If not, you can make copies of the deposit slips in this exercise.

3. Discuss how to endorse a check. When they wish to deposit or cash a check, students will first need to endorse the check. Emphasize the following steps:

   (a) Don't endorse the check until you get to the bank. If it is endorsed and gets lost, someone else could cash it.

   (b) You endorse the check on the back, at the end closest to where it says "Pay to the order of."

   (c) Always sign your name exactly as it appears on the front of the check. If the check is made out to "P.J. Jordan," don't sign it "Patricia Jordan."

   (d) If you wish to deposit the check in your account, write "For Deposit Only" on the back, then endorse it. Do not write "For Deposit Only" if you wish to cash the check.

4. Discuss how to cash a check. Not every bank will cash your check. You may need to go to the bank or credit union where you usually do business or where you have an account. The bank will probably ask to see ID. A good type of identification is a driver's license because it has your picture and signature. Other forms of ID may include a credit card, student ID card, identification badge from where you work, etc.

Activity 12: Writing Checks and Deposit Slips I

Activity 13: Writing Checks and Deposit Slips II

1. If students need more practice, obtain checks and deposit slips from a local bank. Alternately, make copies of the checks and deposit slips provided in Activities 12 and 13, using different amounts for deposits and checks.

2. Review and practice spelling numbers correctly. Many students have difficulty with this!

Activity 14: A Checkbook Register

1. Make up some blank "registers" and give students checks and deposits to record.

2. Show students how to balance a checkbook, using a bank statement and canceled checks.

3. Discuss why it is important to keep a checkbook balanced: to check for errors, to know how much money you have, and to avoid writing overdrafts.

4. Discuss the monthly bank statement. Show students how to read it and talk about why it is important.

5. Explain to the class that some checks come with stubs or with "carbon" duplicates instead of a checkbook register. When you write a check, you fill in the stub at the same time; the duplicate copy type is already done when the original check is written. You also subtract the amount of the check so you can see your balance.

6. Discuss with the class the concept of overdrawing a checking account. If you write a check without having enough money in your checking account to cover it, the check will "bounce," that is, be returned to the store for "insufficient funds." The store will assess a returned-check fee, which may be $25 or more. This is not only an unnecessary expense, but if you write bad checks, stores will no longer accept your checks.

Activity 15: Checking Account Review

Activity 15 can be used as a review or quiz. It includes questions on each topic covered in Activities 7–14.

Unit 3: Saving Your Money (A Savings Account)

To find more information on the Internet about saving accounts, enter the following search terms into a search engine: *savings account rates, bank savings account, banks, credit unions.*

Activity 16: Savings Account Vocabulary

1. Give a spelling test on the six vocabulary words and phrases.
2. Have students write an original paragraph using all six terms.
3. For more advanced students, you may wish to include some of these additional vocabulary words: *joint account, individual account, certificate of deposit (CD), interest compounded daily (or quarterly), principal, percent.*

Activity 17: Opening a Savings Account

1. Discuss why it is best to keep savings in a bank or credit union rather than at home.
2. Most banks have pamphlets available that explain the different types of savings accounts they offer. Obtain several of these. Discuss the features of various types of accounts. Students may decide which accounts would best suit their needs at this time.
3. Ask a bank or credit union officer to come to your class to discuss the various types of savings accounts and certificates of deposit available, as well as the procedures for opening a savings account.
4. Have students learn to compute simple interest. Students could determine how much interest they would earn in one year on $1,000 at various interest rates. The formula $I = P \times R \times T$ may be introduced. (I = interest; P = principal; R = rate; and T = time.) Thus, at 6% interest they would earn $1,000 \times .06 \times 1$ (year) = $60.
5. Discuss other ways to save money besides savings accounts. These might include U.S. Government savings bonds, annuities, Christmas clubs, insurance, or investing in stocks, bonds, or mutual funds.

6. Discuss the concept of compound interest. Most banks compound interest daily, quarterly, or semiannually. In compound interest, the interest is figured on both the principal and the interest the principal has earned. Savings grow faster when interest is compounded more frequently.

7. Discuss the advantages and disadvantages of time deposits, such as certificates of deposit. While they earn more money, the money must be left in the bank for a specified period of time. If the money is withdrawn early, a penalty will be incurred.

Activity 18: Depositing Money in a Savings Account

1. Discuss the purpose of deposit slips. How long and where should they be kept?

2. You may be able to obtain more deposit slips from a bank or credit union to provide additional practice. Or, make additional copies of the deposit slips in Activity 18 and use different deposit amounts for practice.

Activity 19: Withdrawal Slips

1. Discuss the Federal Deposit Insurance Corporation (FDIC). It was formed during the Depression (after a series of bank failures in the 1930s) to protect bank customers. Explain that the FDIC guarantees you will be able to withdraw your money at any time you want it. Individual accounts are insured for up to $100,000.

2. Have students check with a variety of local banks and credit unions to see if they are federally insured. Have them determine how much accounts are insured for.

Activity 20: A Savings Account Passbook

1. For additional practice, make up similar forms on which students can figure the savings account balance.

2. Make up sample passbooks in which "errors" have been made. Have students locate and correct the errors.

Activity 21: Planning Your Spending to Reach Your Goals

1. Discuss the benefits of payroll savings plans.

2. Discuss why it is a good idea to "pay yourself first."

3. It is a good idea to save regularly so you have money to meet unexpected emergencies. Discuss what some of these emergencies might be—e.g., car repairs, illness, lost job, etc.

4. Have students each write a paragraph explaining a goal they would like to save money toward. They should give the cost of the item, how much they are saving weekly, and how long they expect to save before reaching their goal.

5. Discuss the concept of wants versus needs. Have students make a list of things teenagers like to spend their money on—e.g., movies, CDs, lunch money, gas, etc. Then

discuss whether each is a want or a need. The answers may vary somewhat; what is considered a want by one person may be a need for another.

6. Have students each write a paragraph telling how budgeting can help you get the most out of your money.

7. Have students write budgets for their own spending. Or, have them write fictitious budgets.

8. Have students examine the two budgets outlined in Activity 21. What problem do they think might arise with Oscar's budget?

Activity 22: Savings Build-Up

Construct additional problems to figure savings accrued over time.

Activity 23: Savings Account Review

This exercise can be a review or quiz. It includes questions on each topic covered in Activities 16–22.

Unit 4: Automating Your Money (Electronic Banking)

To find more information on the Internet about electronic banking, enter the following search terms into a search engine: *electronic banking, on-line banking, banking on the WWW, banks on the Web, ATM locators.*

Activity 24: Electronic Banking Vocabulary

1. Have a spelling test on the five vocabulary words and phrases.

2. Have students write an original paragraph using all five terms.

3. For more advanced students, you may wish to include some of these terms: *modem, access, verify, electronic, online service, browser, software.*

Activity 25: Using an Automated Teller Machine (ATM)

1. Gather literature from several local banks explaining how to obtain and use an ATM card.

2. As a class, make a list of locations and types of ATMs in your area.

3. Determine the fees charged to use an ATM. The charge will probably be small at the bank with which one has the card. The charge may be much more at another type of ATM.

4. Have students figure the cost to use various types of ATMs different numbers of times per month. For example, if the cost per transaction is $2.50 and you use the ATM five times a month, the total that month would be $12.50. Discuss how the charges can add up if the ATM is used often.

Activity 26: ATM Safety

1. List other precautions students can think of.
2. Discuss any instances of robberies of ATM customers that have taken place in your area.
3. Discuss good places to keep your PIN number.

Activity 27: Debit Cards

1. Get information from several banks about the debit cards they offer. Make a chart on the board to compare card options. Include:

   How do you get a card?
   Is there a cost to use the card?
   Where can it be used?
   When is the purchase subtracted from your account?
   What happens if the card is lost, stolen, etc.?

2. Discuss whether students would prefer to use a checkbook, a debit card, or both.

Activity 28: Banking by Telephone

1. Brochures are available at your local banks that describe their phone banking services. Obtain several and discuss how the services works. What services are offered, and what are the monthly fees? Does the service vary widely from bank to bank?
2. Invite a bank or credit union official to visit your class to discuss the electronic banking services available. Discuss the pros and cons of young adults using these services.

Activity 29: Banking by Computer

1. Obtain literature from one or more local banks that explains how their computer banking is set up. Discuss the features of computer banking as described.
2. Discuss the pros and cons of computer banking as compared with telephone banking.
3. Find out how your bank account information is safeguarded on the Internet.

Activity 30: Electronic Banking Review

Exercise 30 can be used as a review or quiz. It includes questions on each topic covered in Activities 24–29.

Unit 5: Managing Your Money (Credit and Loans)

To find more information on the Internet about credit, enter the following search terms into a search engine: *credit advice, credit cards, credit and loans, loan calculators, credit reports, credit repair, debt management.*

Activity 31: Credit Vocabulary

1. Give a spelling test on the six vocabulary words and phrases.
2. Have students write an original paragraph using all six terms.

3. For more advanced students, you may wish to include these additional vocabulary words: *promissory note, creditor, debtor, co-signer, installment buying, collateral, annual interest rate, credit rating, loan shark, impulse buying,* and *bankruptcy.*

Activity 32: What Is Credit?

1. As a class, make a list of more good and poor reasons for using credit.
2. Discuss how credit cards can make traveling on vacations easier.
3. Discuss how stores hope to make more money by giving their customers credit cards.
4. As a class, make a list of various credit cards that are available.
5. Have students interview a parent or another adult to find out why they do or do not buy on credit and if they have charge accounts or credit cards.
6. Discuss emergencies in which a credit card could be helpful.
7. Discuss the concept of a "credit rating" with students. Explain why it is important to have a good financial reputation.

Activity 33: How Credit Cards Work

Activity 34: Getting a Credit Card

1. Obtain an application for a credit card. Discuss the qualifications stores look for when screening an applicant for a credit card.
2. Have students collect and bring to class examples of credit-card offers. Make a chart to compare these on the board. Features to compare include:

    How much is the annual fee?
    Is there a grace period before interest charges start?
    Is there a monthly usage fee?
    What is the annual percentage rate of interest?
    Is there a charge for late payments?
    Is there a transaction fee for cash advances?
    What is the interest rate for cash advances?
    What is the credit limit?
    Is there a transaction fee for each purchase?

3. Invite a bank officer to talk to the class about establishing good credit as a young adult.
4. Discuss layaway programs, which some stores operate as a form of credit. (The person makes a down payment on an item. Regular payments are made until the item is paid off. The store keeps the item "laid away" until it is fully paid for.)
5. Discuss having a co-signer on a credit-card application or loan. The co-signer agrees to guarantee that you will pay your bill. If you don't, the co-signer must pay. Discuss why you should not co-sign for anyone else.

Activity 35: Reading a Credit-Card Statement

1. Discuss how credit-card companies make a profit. (They charge interest on the unpaid balance each month.)

2. Discuss the importance of checking each charge on the statement carefully. Receipts from each purchase charged should be kept until the statement is received. That way you can check to be sure each charge is correct.

3. Discuss what to do if you find an error on the statement. (Call the toll-free number and talk to a service representative about the problem.)

Activity 36: Getting a Loan

1. Discuss how you might comparison shop for a loan in your town. Where would be the best (and worst) places to look?

2. Have each student telephone one (different) lender to find out their interest rate for a car loan.

3. Discuss the difference between a secured and an unsecured loan. A secured loan is guaranteed by a savings account or property. Two examples are car loans and mortgages. Unsecured loans (or signature loans) require no collateral.

4. Discuss the consequences of having property repossessed. (It is a bad mark on your credit history.)

5. Have a bank or credit union officer explain the procedure for getting a loan. How do you qualify for a loan? Why might teenagers have difficulty obtaining a loan?

6. Discuss the characteristics of a person to whom the bank likes to loan money (a good credit risk). This person is over 21, makes a good salary, has been at the same job for more than two years, has lived at the same address for more than two to three years, has a bank account, pays bills on time, and is borrowing no more than two months' salary. Discuss why each of these characteristics is important to a lender.

Activity 37: Paying Off a Loan

1. Discuss what recourse the lender has if you fail to meet your loan payments. (Repossession.)

2. Discuss the time factor in paying off a loan. A person taking twice as long to pay off a loan will pay much more interest.

3. Discuss why it is not advisable to take out a five-year car loan in most cases. (The car may not be running in five years, but you will still be paying for it. The total cost of the car will be much more this way.)

4. Ask a lender who deals with car loans to come to class and discuss obtaining a loan with favorable terms.

Activity 38: How to Save Money on Loans

1. Decide as a class on a situation requiring a loan—for example, buying a car). Assign students to find the cost of the loan at different places. Make a chart to show your findings.

2. Discuss the "Truth in Lending" law. This law requires the lender to tell the total finance charge on a loan.

3. Have students figure the amount of interest that will be paid at different rates of interest.

Activity 39: Making a Down Payment

1. Have students figure the finance charge at 18 percent a year on an item purchased with several different down payments.
2. Have students collect ads stating that a down payment is required. Figure out how much this would be. (For example, if 10 percent down is required on a $600 TV, the down payment must be at least $60.)

Activity 40: Using Credit Responsibly

1. Have students write a paragraph agreeing or disagreeing with the adage: "Neither a borrower nor a lender be."
2. Ask a credit manager from a local store to talk to your class on common problems with credit, especially those affecting young people. Other subjects should include how students can avoid those problems and what steps the store takes if you don't make your payments.
3. Discuss how it could happen that a wealthy person could get into trouble with debt.
4. Discuss why once you get behind on your bills, it is so hard to catch up.
5. Have students list bills a young person living in a first apartment might have. How would credit-card or loan debt fit into that picture?
6. Discuss what might happen if you got into money trouble. (Repossession of property, bad checks, poor credit rating, bankruptcy.)
7. Discuss how free or low-cost credit counseling is available for those with money troubles.

Activity 41: Credit Review

This exercise can be used as a quiz or review on the topics covered in Activities 31–40.

Name _____ Date _____

Paycheck Vocabulary

Hourly pay	How much a person earns for every hour worked (per hour).
Gross pay	Hourly pay multiplied by the number of hours worked. Gross pay is how much is actually earned before deductions are taken out of the paycheck.
Deduction	Money taken out of a person's paycheck. Some deductions might be: insurance, income tax, Social Security (FICA), union dues, or money put in a savings account.
Net pay	How much money a person takes home after deductions.

◆ **Fill in the Blanks.** Fill in each blank using a word or phrase from the box above. You will use each term only once.

1. Take-home pay is called _____ .

2. Amounts of money that are subtracted from a person's gross pay are called

   _____ .

3. If you earn $8.50 an hour, your _____ is $8.50.

4. The money a person earns before any deductions are taken out is called

   _____ .

◆ **Matching.** Match each item in the left column with a word or phrase in the right column. Write the letter of the correct item in each blank on the left. You will use some choices in the right column more than once.

5. _____ income tax   **(a)** hourly pay

6. _____ take-home pay   **(b)** gross pay

7. _____ union dues   **(c)** deduction

8. _____ earning per hour   **(d)** net pay

9. _____ pay before deductions

10. _____ Social Security (FICA)

11. _____ insurance

12. _____ money put in savings

Name _____ Date _____

Finding Number of Hours Worked

> Joey had a part-time job. He knew he should keep track of the hours he worked. That way he could be sure his paycheck was right.
>
> Joey worked 2 hours Monday through Friday after school. On Saturday he worked 8 hours. He wanted to know how many hours he had worked.
>
> First Joey multiplied 2 hours by 5 days (Monday through Friday). This made 10 hours worked.
>
> Next Joey added the 8 hours he worked on Saturday to the 10 hours he had worked during the week. This made a total of 18 hours worked.

◆ **Directions:** Find how many hours each person below worked.

1. Teresa worked from 1:00 P.M. to 4:00 P.M. Monday through Friday.	**4.** Leo worked 3 hours each on Tuesday and Thursday. On Friday, he worked from 3:00 P.M. to 10:00 P.M.
2. Sara worked from 4:00 P.M. to 6:00 P.M. on Monday, Wednesday, and Friday.	**5.** Etienne worked from noon until 6:00 P.M. Monday through Friday. On Saturday, he worked 8 hours.
3. David worked from 8:00 a.m. to 4:00 p.m. Monday through Friday. On Saturday, he worked from 10:00 A.M. to 6:00 P.M.	**6.** Chandra worked 2 hours a day, Monday through Friday. On Saturday, she worked 4 hours.

Name _____ Date _____

Finding Gross Pay

> Mark's part-time job paid $5.50 an hour. One week, he worked 20 hours. Mark wanted to find his gross pay.
>
> To find his gross pay, Mark multiplied the number of hours he worked (20) by his hourly pay ($5.50).
>
> $$\$5.50 \times 20 = \$110.00$$
>
> Mark's gross pay was $110 that week.
>
> ```
>    5.50
>  x  20
>  ------
>    000
>   1100
>  ------
>  110.00
> ```

◆ **Directions:** Find each person's gross pay:

1. Elena worked for 20 hours at Jill's Boutique. She earned $5.15 an hour.	**5.** Jamal worked for 18 hours at the movie theater. He earned $5.25 an hour.
2. Sara worked for 10 hours at Joe's Burgers. She earned $4.85 an hour.	**6.** Jose worked for 12 hours at the ice cream shop. He earned $5.85 an hour.
3. Tia baby-sat for 14 hours one weekend. She earned $4.25 an hour.	**7.** Martin worked for 15 hours at Skateland. He earned $5.17 an hour.
4. Tran worked for 25 hours on his uncle's farm. He earned $6.00 an hour.	**8.** Keisha worked for 20 hours at the grocery store. She earned $5.95 an hour.

Name _____ Date _____

Adding Deductions

Sumida had a job in a factory. The hourly pay was $6. She worked 40 hours one week. Sumida figured her gross pay would be $240 ($6 × 40 hours). But when she got her paycheck, it was only $159. What went wrong?

Sumida forgot some deductions would be taken out of her gross pay. Her deductions were:

Federal Income Tax	$36	State Income Tax	$15
FICA (Social Security)	$16	Health Insurance	$3
Retirement	$3	Union Dues	$3
Savings Plan	$5		

To find her total deductions, add all the deductions up. Sumida had $81 in deductions. ($36 + $16 + $3 + $5 + $15 + $3 +$3)

◆ **Directions:** Find the total deductions on each paycheck stub. Write the total in the box marked "Total Deductions."

Name: **Kim Chang**		Date of Check: **May**		
Federal Withholding Tax $57.29		Social Security $25.15	Retirement	State Income Tax $6.85
Health Insurance	Savings $10.00	Life Insurance	Parking	Other
Total Deductions				

Name: **LaToya Miller**		Date of Check: **July**		
Federal Withholding Tax $89.23		Social Security $45.19	Retirement $38.00	State Income Tax $10.50
Health Insurance $13.95	Savings $50.00	Life Insurance $2.98	Parking $6.00	Other
Total Deductions				

Name: **Sven Paulsen**		Date of Check: **September**		
Federal Withholding Tax $159.92		Social Security $72.93	Retirement $64.00	State Income Tax $17.97
Health Insurance $27.50	Savings $5.19	Life Insurance	Parking	Other
Total Deductions				

       41 Activities in Basic Money Management

Name _____ Date _____

Finding Net Pay

> It is a good idea to know how to check a paycheck to make sure that the net pay has been figured correctly.
>
> To find the net pay:
>
> **1.** Add all the deductions.
>
> **2.** Subtract the total deductions from the gross pay.

◆ **Directions:** Find the total deductions and the net pay in each example below.

1.

Gross Pay		179.95
Deductions:		
Federal Withholding	35.19	
State Income Tax	4.20	
FICA	17.50	
Health Insurance	7.89	
Dues	4.50	
Savings	15.00	
Total Deductions		
Net Pay		

3.

Gross Pay		235.00
Deductions:		
Federal Withholding	45.19	
State Income Tax	3.58	
FICA	20.45	
Health Insurance		
Dues		
Savings	15.00	
Total Deductions		
Net Pay		

2.

Gross Pay		589.19
Deductions:		
Federal Withholding	112.56	
State Income Tax	12.09	
FICA	68.14	
Health Insurance	13.50	
Dues	5.19	
Savings	50.00	
Total Deductions		
Net Pay		

4.

Gross Pay		785.09
Deductions:		
Federal Withholding	175.14	
State Income Tax	20.45	
FICA	78.40	
Health Insurance	25.19	
Dues		
Savings	55.00	
Total Deductions		
Net Pay		

Name _____ Date _____

Paycheck Review

---

1. Joan has a part-time job at the mall. She works 4 hours every Monday, Wednesday, and Friday. On Saturday she works 8 hours. What are her total hours per week?

3. Carlos's paycheck deductions are:

Federal Income Tax: $35.50
State Income Tax: $2.05
FICA: $5.00
Insurance: $1.29

What are Carlos's total deductions?

---

2. Tony works 20 hours per week. He earns $5.15 per hour. What is his gross pay?

4. Complete Shakira's paycheck stub using the information below. What is her net pay?

Gross pay: $312.00
Deductions:
   Federal Income Tax: $35.10
   State Income Tax: $3.01
   FICA: $7.50
   Insurance: $2.19

> ### Shakira Roy
> Gross Pay _____
> Total Deductions _____
> Net Pay _____

◆ **Matching.** Match each item in the left column with a definition in the right column. Write the letter of the correct definition in each blank on the left. Use each definition only once.

5. _____ deduction
6. _____ gross pay
7. _____ hourly pay
8. _____ net pay
9. _____ FICA

(a) take-home pay
(b) what a person earns in 1 hour
(c) money taken out of a person's pay for taxes, insurance, or dues
(d) a person's earnings before deductions are taken out
(e) Social Security

41 Activities in Basic Money Management

Name _____ Date _____

Checking Account Vocabulary

Check	An order to pay money from your checking account to a person or company.
Deposit	Putting money into a bank account.
Deposit slip	A paper you fill out to show how much money (cash or checks) you are depositing.
Signature	The way you sign your name.
Checkbook register	A record of checks and deposits made to a checking account.
Balance	How much money is in the account.

◆ **Fill in the Blanks.** Fill in each blank using a word or phrase from the box above. You will use each term only once.

1. If Vic has $14.98 in his checking account, then his _____ is $14.98.

2. If Nicki put $50 into her checking account, she made a _____ .

3. All checks should be recorded in the _____ .

4. When making a deposit, Ramon gave the teller the money and a _____ .

5. At the grocery store Mrs. Alvarez wrote a _____ for $48 worth of food.

6. A check is not good without your _____ on the bottom line.

◆ **Matching.** Match each item in the left column with a word or phrase in the right column. Write the letter of the correct item in each blank on the left. Use each choice only once.

7. _____ a paper that shows money being deposited   **(a)** check

8. _____ how you sign your name   **(b)** deposit

9. _____ record of checks   **(c)** deposit slip

10. _____ an order to pay money   **(d)** signature

11. _____ money now in an account   **(e)** checkbook register

12. _____ putting money in an account   **(f)** balance

Name _____ Date _____

Opening a Checking Account

Choose a bank or credit union that is convenient and offers the services you want. You must go to the bank in person to open an account. Go to the "New Accounts" desk. Ask what kinds of checking accounts the bank offers. Choose the one you want.

You will be asked to fill out a signature card. This card gives the bank a record of how you sign your name. You should sign every check you write the same way you signed the signature card.

Next you will need to deposit some money in your account. You will also choose the kind of checks you want. They will have your name and address printed on them.

◆ **Directions:** Fill out the sample signature card below. Use your own name and address and today's date. Your account number is 30481.

Date _____	Account No. _____

Signature Card For:

Name(s): _____

All checks require _____ of the _____ authorized signatures herein.

The undersigned certify that the signatures on this card are the duly authorized signatures which you will recognize on checks, drafts, money orders, and other instruments drawn by us against our Deposit Account No. _____.

Title	Type Name Below		Sign Name Below
_____	_____	will sign	_____
_____	_____	will sign	_____
_____	_____	will sign	_____

Address _____

1. How do you think the signature card protects the checking account owner?

   _____

   _____

2. Why do you think there is space on the card for more than one signature?

   _____

   _____

Name _____ Date _____

Writing Checks

It is important to write checks correctly. Here's how:

1. Always write in pen, never in pencil.
2. Use today's date.
3. On the line that says "Pay to the order of . . ." write the name of the person or business to whom you are writing the check.
4. Write the amount of money twice: in numbers (for example, "$14.98") and in words (for example, "Fourteen dollars and 98/100").
5. If you want to, write what the check is for on the line marked "For."
6. Sign your name on the line at the bottom right.

◆ **Directions:** Study the sample check below. Then fill in the check at the bottom of the page.

Example:
A check for
$23.98 is written
to LaToya Williams.

John P. Customer	85-590
521 Money Street	653
Riches, MS 39211	643

Date *10-2-99*

Pay To The Order Of *La Toya Williams* $ *23 98/100*

Twenty-three and 98/100 Dollars

UNION BANK

For *club dues* *John P. Customer* MP

⑆065306532⑆0157 ⑈5081522⑈ 643

Write this check
to Food Center
for $117.63.
Use today's date.
Sign your own name.

Your Own Name	85-590
Your Street	653
Your City, State, Zip	644

Date _____

Pay To The Order Of _____ $ _____

_____ Dollars

UNION BANK

For _____ _____ MP

⑆065306532⑆0157 ⑈5081522⑈ 644

Name _____ Date _____

Writing More Checks

◆ **Directions:** Fill in the checks below using the information given. Use today's date. Use your own signature.

A check for
$17.97 to
The Bluz Music Shop

Your Own Name
Your Street
Your City, State, Zip
85-590
653
645

Date _____

Pay To
The Order Of _____ $ _____

_____ Dollars

UNION BANK

For _____ _____ MP

⑈:065306532⑈:0157 ⑈5081522⑈ 645

A check for
$129.76
to McDill's Dept. Store

Your Own Name
Your Street
Your City, State, Zip
85-590
653
646

Date _____

Pay To
The Order Of _____ $ _____

_____ Dollars

UNION BANK

For _____ _____ MP

⑈:065306532⑈:0157 ⑈5081522⑈ 646

A check for
$57.20 to
Valley Electric

Your Own Name
Your Street
Your City, State, Zip
85-590
653
647

Date _____

Pay To
The Order Of _____ $ _____

_____ Dollars

UNION BANK

For _____ _____ MP

⑈:065306532⑈:0157 ⑈5081522⑈ 647

Name _____ Date _____

Making Deposits to a Checking Account

When you want to deposit money in your checking account, you must fill out a deposit slip. Here's how:

1. Fill in the date.

2. If you are depositing cash, write the total amount of cash being deposited where it says "cash."

3. If you are depositing checks, list them one by one on the lines where it says "checks."

4. If you want to keep some of the money, write the amount you want where it says "Less cash received." When you have received the correct amount back, sign the slip on the line marked "Sign here for cash received."

5. Write the amount of the total deposit.

◆ **Directions:** Study the sample deposit slip below. Then fill out the deposit slip at the bottom of the page, using today's date.

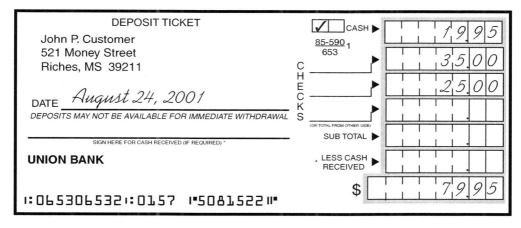

You deposit:
$45.00 in bills
$.75 in coins
A check for $100.

Name _____ Date _____

Writing Checks and Deposit Slips I

◆ **Directions:** Using the amounts of money given, complete the checks and deposit slip below. Use today's date and your own signature.

1. Deposit $35.00 in cash and checks for $11.95, $35.80, and $139.92.

DEPOSIT TICKET

✓ CASH ▶

Your Own Name
Your Street
Your City, State, Zip

85-590 1
653

C
H
E
C
K
S

DATE _____
DEPOSITS MAY NOT BE AVAILABLE FOR IMMEDIATE WITHDRAWAL

(OR TOTAL FROM OTHER SIDE)

SUB TOTAL ▶

_____
SIGN HERE FOR CASH RECEIVED (IF REQUIRED) *

UNION BANK

. LESS CASH
RECEIVED

$

⑈065306532⑈:0157 ⑈5081522⑈

2. Write check #648 to Gomez Auto Parts for $49.85.

Your Own Name
Your Street
Your City, State, Zip

85-590
653

648

Date _____

Pay To
The Order Of _____ $_____

_____ Dollars

UNION BANK

For_____   _____ MP

⑈065306532⑈:0157 ⑈5081522⑈ 648

3. Write check #649 to Musicland Records for $25.90.

Your Own Name
Your Street
Your City, State, Zip

85-590
653

649

Date _____

Pay To
The Order Of _____ $_____

_____ Dollars

UNION BANK

For_____   _____ MP

⑈065306532⑈:0157 ⑈5081522⑈ 649

Name _____ Date _____

Writing Checks and Deposit Slips II

◆ **Directions:** Using the amounts of money given, complete the checks and deposit
slip below. Use today's date and your own signature.

1. Write check #650 to Siempre Bella Clothing for $75.18.

```
Your Own Name                              85-590              650
Your Street                                 653
Your City, State, Zip
                              Date _____

Pay To
The Order Of _____| $_____

_____ Dollars

UNION BANK

For_____                _____  MP

I:0653065321:0157  I·5081522II·  650
```

2. Deposit checks for $67.98, $54.64, and $15.30. Keep $25.00 in cash.

```
                    DEPOSIT TICKET                    ✓  CASH ▶
        Your Own Name                            85-590  1
        Your Street                               653
        Your City, State, Zip                  C
                                               H
        DATE _____           E
        DEPOSITS MAY NOT BE AVAILABLE FOR      C
        IMMEDIATE WITHDRAWAL                   K
                                               S   (OR TOTAL FROM OTHER SIDE)
        _____            SUB TOTAL ▶
        SIGN HERE FOR CASH RECEIVED (IF REQUIRED)
                                                    LESS CASH ▶
        UNION BANK                                  RECEIVED

                                                  $
        I:0653065321:0157  I·5081522II·
```

3. Write check #651 to Rico's Sport Shop for $35.90.

```
Your Own Name                              85-590              651
Your Street                                 653
Your City, State, Zip
                              Date _____

Pay To
The Order Of _____| $_____

_____ Dollars

UNION BANK

For_____                _____  MP

I:0653065321:0157  I·5081522II·  651
```

Name _____ Date _____

A Checkbook Register

A checkbook register is where you keep track of checks you write and deposits you make to your checking account. Each check and deposit must be recorded in the register. By doing so, you can see how much money is in your account.

Each time you write a check, write it in the register. Include:
- the check number
- the date you wrote the check
- whom the check was written to
- what the check was for (food, jeans, etc.)
- how much the check was for

When you make a deposit, write it in the register. Include:
- the date you made the deposit
- where the deposit came from (e.g., Oct. paycheck)
- the amount deposited

◆ **Directions:** Complete the register below, using the information given. Find the balance after each item. The first one is done for you.

A. 3/2   Check #819 to DiPietro's for food ($55.50)
B. 3/5   Paycheck for 2/17 to 3/3 ($459)
C. 3/7   Check #820 to Valley Gas ($17.89)
D. 3/8   Check #821 to Bell Telephone ($32.98)
E. 3/10   Check #822 to Terrace Apartments for rent ($350)
F. 3/20   Check #823 to Mark Toyota for car payment ($159)
G. 3/19   Paycheck for 3/4 to 3/18 ($459)
H. 3/20   Check #824 to DiPietro's for food ($65.00)
I. 3/25   Check #825 to Safety Insurance for car insurance ($49)

NUMBER	DATE	CODE	DESCRIPTION OF TRANSACTION	PAYMENT/DEBIT (−)	FEE (−)	TAX	DEPOSIT/CREDIT (+)	BALANCE $ 200 00
819	3/2		DiPietro's	$ 55 50			$	− 55 50
								144 50

Name _____ Date _____

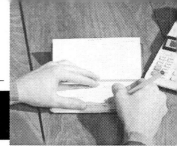

Checking Account Review

◆ **Directions.** Fill in the deposit slip and check below. Use your own name and address and today's date.

1. Deposit:
Cash: $25.00
Checks: $ 7.85
$ 8.17
$ 5.04

DEPOSIT TICKET
Your Own Name
Your Street
Your City, State, Zip

☑ ☐ CASH ▶
85-590 1
653

C
H
E
C
K
S

DATE _____
DEPOSITS MAY NOT BE AVAILABLE FOR IMMEDIATE WITHDRAWAL

(OR TOTAL FROM OTHER SIDE)
SUB TOTAL

SIGN HERE FOR CASH RECEIVED (IF REQUIRED) *

UNION BANK

* LESS CASH
RECEIVED

$

⑈065306532⑈0157 ⑈5081522⑈

2. Write check #764 to Rialto Hardware for $35.08.

Your Own Name
Your Street
Your City, State, Zip

85-590
653

764

Date _____

Pay To
The Order Of _____ $ _____

_____ Dollars

UNION BANK

For_____ _____ MP

⑈065306532⑈0157 ⑈5081522⑈ 764

◆ **Word Problems.** Find the balance in each checking account below.

3. Frank had $75.00 in his checking account. He wrote checks for $14.50, $7.04, and $18.95. What was his balance? _____

4. Dan had $100.00 in his checking account. What was his balance after he wrote checks for $75.03 and $19.80? _____

◆ **Matching.** Match each item in the left column with a definition in the right column. Write the letter of the correct definition in each blank on the left. Use each definition only once.

5. _____ check
6. _____ deposit
7. _____ deposit slip
8. _____ signature
9. _____ checkbook register
10. _____ balance

(a) how you sign your name
(b) record of checks and deposits
(c) an order to pay money
(d) total money now in an account
(e) putting money in an account
(f) list of money being deposited

Name _____ Date _____

Savings Account Vocabulary

Budget	A plan for spending one's money.
Withdrawal	Money taken out of a savings account.
Passbook	A book for keeping a record of deposits, withdrawals, and the balance in a savings account.
Interest	Money the bank pays you for keeping your money in a savings account.
Deposit slip	A bank paper that says how much money you deposited (put into your bank account).
Account number	The number that identifies your account.

◆ **Fill in the Blanks.** Fill in the blanks using each word or phrase in the box above only once.

1. Jean kept $500 in a savings account for 1 year. At the end of the year she had $525.

   The extra $25 is _____ .

2. The number 04139 stamped on George's passbook is his _____ .

3. The bank's computer prints out the deposits, withdrawals, and current balance in

   Ming-na's _____ .

4. Tony took $20 out of his account. He made a _____ .

5. Layla wrote down a plan for spending and saving her money. Layla's plan is her

   _____ .

6. The teller gave Sergei a copy of the _____ showing the

   amount of money had just deposited.

◆ **Matching.** Match each item in the left column with a word or phrase in the right column. Write the letter of the correct term in each blank on the left. Use each word or phrase only once.

7. _____ the number that identifies your account   **(a)** budget

8. _____ money earned by your money   **(b)** withdrawal

9. _____ a book showing deposits and withdrawals   **(c)** passbook

10. _____ a bank paper you get when you make a deposit   **(d)** interest

11. _____ a money plan   **(e)** deposit slip

12. _____ taking money out of an account   **(f)** account number

Name _____ Date _____

Opening a Savings Account

There are many places people save money: in a piggy bank, under a mattress, buried in the garden, or in a book, just to name a few!

A **savings account** is a better way to save for several reasons:

1. Your money is insured against loss.
2. You will earn interest on the money.
3. The money will not be accidentally lost or stolen.

When you are ready to open a savings account, think carefully about which bank or credit union you will choose. You will want a bank that is convenient to where you live and work. Here are a few questions to ask the banks you are thinking about:

1. Are you federally insured?
2. What is the minimum amount of money needed to open an account?
3. What types of accounts are available?
4. What interest rate will be paid?
5. How often will interest be paid?
6. Are there service charges on the account?
7. What are the bank's hours?

◆ **Directions:** Answer each question briefly.

1. Give two reasons for putting your money in a bank or credit union.

   _____

   _____

2. Why do you think it can be a good idea to check more than one bank or credit union before you open your savings account?

   _____

   _____

3. Why do you think it is wise to avoid savings accounts that charge a monthly service charge?

   _____

   _____

Name _____ Date _____

Depositing Money in a Savings Account

A savings account works like this: You take your money and passbook to the bank. Then you fill out a deposit slip. You give the deposit slip, the money, and your passbook to a bank teller. The teller will record your deposit in your passbook and give you a copy of the deposit slip to show that the bank received your money.

◆ **Directions:** Fill in the deposit slips below. Your account number is 13429. Use your own name and address.

1. On March 18 of this year, deposit:
   $25.00 in bills (currency)
   $1.25 in coins
   a check for $50

	DOLLARS	CENTS
CURRENCY		
COIN		
CHECKS		
TOTAL		

ACCOUNT NUMBER	TOTAL DEPOSIT

SAVINGS DEPOSIT	DATE

Signature

Street

City & State

2. On August 14 of this year, deposit:
   $32.00 in bills
   $2.19 in coins
   a check for $54.98
   a check for $15

	DOLLARS	CENTS
CURRENCY		
COIN		
CHECKS		
TOTAL		

ACCOUNT NUMBER	TOTAL DEPOSIT

SAVINGS DEPOSIT	DATE

Signature

Street

City & State

3. On April 7 of this year, deposit:
   $18.00 in bills
   $3.79 in coins
   a check for $5
   a check for $7.85
   a check for $14.03

	DOLLARS	CENTS
CURRENCY		
COIN		
CHECKS		
TOTAL		

ACCOUNT NUMBER	TOTAL DEPOSIT

SAVINGS DEPOSIT	DATE

Signature

Street

City & State

Name _____ Date _____

Withdrawal Slips

> Sometimes you'll want to withdraw (take out) money from your savings account. At these times you'll need to have your passbook and a withdrawal slip to give to the teller.

◆ **Directions:** Complete the withdrawal slips below. Fill in your address, the date, your account number, and the amount of money. Then sign the slip. Your account number is 13429. Use today's date.

1. Withdraw $35.

SAVINGS WITHDRAWAL AUTHORIZATION

Date _____

AMOUNT _____

I HEREBY AUTHORIZE THIS AMOUNT CHARGED TO SAVINGS ACCOUNT NO.

_____

_____
Signature

_____
Street

_____
City & State

2. Withdraw $125.

SAVINGS WITHDRAWAL AUTHORIZATION

Date _____

AMOUNT _____

I HEREBY AUTHORIZE THIS AMOUNT CHARGED TO SAVINGS ACCOUNT NO.

_____

_____
Signature

_____
Street

_____
City & State

Name _____ Date _____

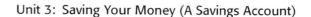

A Savings Account Passbook

The inside of a savings account passbook looks something like the example below. The bank's computer or a teller will record your deposits, withdrawals, and balance in the passbook. (But you should know how to do this yourself to be sure your balance is correct.)

◆ **Directions:** Find the balance after each deposit or withdrawal is made in the passbook below. You should **add** all deposits or interest payments to the balance. **Subtract** all withdrawals from the balance. The first two are done for you.

Date	Amount of Withdrawal	Amount of Deposit	Interest Paid	Balance
9/1		$50.00		$ 50.00
9/15		$50.00		$100.00
10/1		$50.00		
10/15		$55.00		
10/16	$15.00			
10/16			$2.27	
11/1		$55.00		
11/3	$40.00			
11/15		$55.00		
12/1		$55.00		
12/15		$55.00		
12/17	$250.00			
12/29			$2.35	

Name _____ Date _____

Planning Your Spending to Reach Your Goals

You have been learning about savings accounts. Many people feel it is hard to save money. By the end of the month, they find they have no money left to put in savings.

The answer to this problem is **always to pay yourself first.** Every time you get paid, put some of the money in savings first. This way you won't have a chance to spend it.

You may be saving money to reach a goal, such as money for a car. Or, you may wish to save to meet unexpected needs, such as car repairs. Regular saving can help you meet these goals.

Here are a few ways to free up more cash for savings:

1. Save some money from each paycheck. Ask your boss if it can be automatically deducted from your paycheck each month. If you never have it, you won't miss it.

2. Put part (or all) of the money received at holidays into savings.

3. Don't carry much cash. If you have to write a check, you'll be less tempted to spend.

4. Make sure you know the difference between wants and needs. For example, you may **need** a pair of running shoes. But you may **want** a pair that costs much more than you really need.

5. Plan your spending by making a simple budget. Make a list of your needs. Include savings in this list. Subtract these amounts from your income. Is anything left over? If so, that is money to use for extra wants or to save.

◆ **Directions:** Fill in the budget for each person below. Find out how much each person has left to spend after needs and savings have been paid for.

1. Oscar earns $232.90 a month. His needs are $30/month for school lunches. He also spends $24/month for bus fare. Oscar hates riding the bus. He wants to save $150 monthly toward buying a car.
Earnings
Lunches
Bus fare
Savings
Total expenses
Spending money

2. Yolanda earns $319 a month. She needs $30/month for school lunches. She spends $50/month for gas and oil. She wants to save $100/month toward a summer vacation.
Earnings
Lunches
Bus fare
Savings
Total expenses
Spending money

         41 Activities in Basic Money Management

Name _____ Date _____

Savings Build-Up

A small amount saved regularly in a savings account will soon build up. Plus, the bank will add interest to the money you put in the account.

◆ **Directions:** Read the problems below. Decide how much money each person will have at the end of the time given. The first one is done for you.

Example:

Takashi saved $20 a week from his part-time job. How much money did he have after one month?

$20 × 4 = $80

He had $80 (plus interest).

---

1. Jake saved $5 a week.

   How much money did he have:

   after 4 weeks? _____

   after 26 weeks? _____

   after 52 weeks (1 year)? _____

   after 5 years? _____

   *With 6 percent interest added, Jake had $1,554 in 5 years. How much interest had he earned over 5 years? _____

2. Rose's grandmother gave her $100 every year on her birthday and at Christmas. Rose saved the money. How much did she have:

   after 1 year? _____

   after 3 years? _____

   after 5 years? _____

   *With 6 percent interest added, Rose had $1,193. How much interest had she earned over 5 years?

   _____

3. Hector saved $85 a month from his paycheck. How much did he have:

   after 6 months? _____

   after 1 year? _____

   after 5 years? _____

   *With 6 percent interest added, Hector had $6,093. How much interest had he earned over 5 years?

   _____

4. Maria saved $5 a week from her allowance and $25 a week from her part-time job. How much did she have:

   after 1 week? _____

   after 1 year? _____

   after 5 years? _____

   *With 6 percent interest added, Maria had $9,318. How much interest had she earned over 5 years?

   _____

Name _____ Date _____

Savings Account Review

◆ **Directions.** Fill in the deposit slip below. Your account number is 13429. Use today's date and your own name and address.

1. Deposit:
   $20.00 in bills
   (currency)
   $5.08 in coins
   a check for $3.19
   a check for $8.05

	DOLLARS	CENTS
CURRENCY		
COIN		
CHECKS		
TOTAL		

ACCOUNT NUMBER	TOTAL DEPOSIT

SAVINGS DEPOSIT	DATE

Signature

Street

City & State

◆ **Word Problems.** Find the answer to each savings problem below.

2. Eli saved $50 a month from his part-time job. How much had he saved after 1 year?

   _____

3. Martin needed $600 more for a used car he wanted. He saved $150 a month. How long would it take him to save the $600?

   _____

4. Consuelo had $300 in her savings account. During April, she deposited $50 and $65. She also made a $25 withdrawal. What was her balance at the end of the month?

   _____

5. Maya made $200 a month from her part-time job. She needed $20 for lunches, $25 for gas, and $50 for savings. How much was left over for other things? _____

◆ **Matching.** Match each item in the right column with its definition in the left column. Write the letter of the correct word or phrase in each blank on the left. Use each choice only once.

6. _____ the number identifying your account
7. _____ money earned by your money
8. _____ a book listing deposits, withdrawals, and the balance in a savings account
9. _____ a bank paper you get when you deposit money
10. _____ a money plan
11. _____ taking money out of an account

(a) budget
(b) withdrawal
(c) passbook
(d) interest
(e) deposit slip
(f) account number

Name _____ Date _____

Electronic Banking Vocabulary

ATM	Automated teller machine—a cash machine found at banks and other locations.
Debit card	A check card. A debit card works like a check. Money is taken from your checking account to cover each purchase.
PIN	Personal identification number—a secret code you need to be able to use the ATM.
Service charge	Money charged by the bank for a service.
Transaction	Banking business, such as a deposit or withdrawal.

◆ **Fill in the Blanks.** Fill in each blank using a word or phrase from the box above. You will use each term only once.

1. A _____ can be used like a check.

2. There is a _____ for paying bills by phone.

3. To use an ATM, you will need your card and your _____ .

4. To get cash when the bank is closed, you could use the _____ .

5. Making a deposit is an example of a bank _____ .

◆ **Matching.** Match each item in the left column with a word or phrase in the right column. Write the letter of the correct item in each blank on the left. You will use some choices in the right column more than once.

6. _____ a code used in the ATM

7. _____ a fee

8. _____ a deposit or withdrawal

9. _____ a check card

10. _____ cash machine

11. _____ may be found at a store

12. _____ works like a check

13. _____ a secret code

14. _____ money charged by the bank

15. _____ a piece of banking business

(a) debit card

(b) ATM

(c) PIN

(d) transaction

(e) service charge

Name _____ Date _____

Using an Automated Teller Machine (ATM)

> With an ATM, you can bank any time of the night or day. You can deposit or withdraw money from your checking or savings accounts. You can check your account balances.
>
> You get an ATM card from the bank where you have an account. You also have a PIN, a secret code you need to use the machine. Never tell anyone your PIN.
>
> To use your ATM card, place it in the slot in the machine. Enter your PIN. Then answer the questions the machine asks.

◆ **Directions:** You have placed your ATM card in the machine and entered your PIN. Now answer the questions the machine asks. Circle each correct answer choice in the boxes below, or write in the correct amount in the white boxes.

The ATM screen says:	1. You want to withdraw $75 from your checking account.	2. You want to deposit $200 to your savings account.	3. You want to check the balance in your checking account.
• Do you want to make a withdrawal?	yes/no	yes/no	yes/no
• Do you want to make a deposit?	yes/no	yes/no	yes/no
• Do you want to check your balance?	yes/no	yes/no	yes/no
• Which account?	savings/ checking	savings/ checking	savings/ checking
• What amount?			
• Place your deposit in the envelope. Put the envelope in the slot.			
• Take your money.			
• Do you want to make another transaction?	yes/no	yes/no	yes/no
• Take your card.			

         41 Activities in Basic Money Management

Name _____ Date _____

ATM Safety

> There are several things you should do for safety when using an ATM:
> - **Keep your PIN secret.** Don't write it on your ATM card. Don't even carry it in your wallet with your ATM card.
> - **Look around before using the ATM.** If there are people around who make you uncomfortable, don't use it. Find another machine.
> - When using the ATM, **don't let people stand close enough** to see your PIN.
> - **Don't count your money at the ATM.** Take your money, your card, and your receipt back to your car or another safe place.
> - If you use an ATM at night, **choose one that is well-lit.** If you can, **take a friend** with you.
> - **Protect ATM cards and debit cards** as if they were cash.

1. Why do you think you should never write your PIN on your ATM card?

   _____

   _____

   _____

2. What do you think you should do if someone stands too close to you while you are using the ATM? _____

   _____

   _____

3. If you are using a drive-up ATM (from your car), what do you think you might do to ensure your safety? _____

   _____

   _____

4. Why do you think you should protect your debit card or ATM card as if it were cash?

   _____

   _____

   _____

Name _____ Date _____

Debit Cards

> What looks like a credit card but is not a credit card? Answer: a debit card!
>
> The debit card is also called a check card. It looks just like a major credit card. It may even have the VISA or Mastercard logo on it.
>
> If you pay for a purchase with a debit card, it is just like using a check. The amount you spend will be taken out of your checking account. Debit cards can also be used in ATMs.
>
> Be sure to write down what you spend with your debit card. Subtract it from the balance in your checking account.

1. Why do you think some people might prefer using a debit card instead of writing checks for purchases? _____

_____

_____

2. Why do you think you should record all uses of the debit card in your checkbook register?

_____

_____

_____

3. How is a debit card different from a credit card? _____

_____

_____

4. Do you think it would be easier to overspend by using a debit card than by writing checks? Why or why not? _____

_____

_____

Name _____ Date _____

Banking by Telephone

Telephone banking lets you pay bills and do other banking business by phone 24 hours a day. Here's how it works:

- Many banks will let you check your balance, find out which checks or deposits have cleared, or transfer money between accounts with no service charge. Get the telephone number from the Yellow Pages or from the bank. Just dial the number and follow the instructions. (You will need a touch-tone phone.)
- To pay bills by phone, you must fill out a special application with the bank. You must tell them which bills you will be paying by phone. The bank will give you a special code for each company you wish to pay by phone. (The service works best for bills you pay every month.)

Once you have this service set up, it is easy to pay bills. You pick up the phone and key in the code for the company you want to pay. Then you key in the payment date and the amount.

There may be a monthly service charge to pay your bills by phone.

1. Why do you think it might be helpful to be able to check the balance in your account by

   telephone? _____

   _____

2. How could you tell if using a bill-paying account would be worth a $4 monthly service fee?

   _____

   _____

3. Why do you think a bill-paying account might save you time? _____

   _____

   _____

4. Why might you want to find out if a check has cleared (been subtracted from your account)?

   _____

   _____

Name _____ Date _____

Banking by Computer

> Many banks will allow you to do your banking business using a computer and the Internet. To do this, you will need a computer, a modem, and special software (a computer program) that you can get from the bank.
>
> This will allow you to check your account balances, transfer money between accounts, check interest rates, pay bills, and much more.
>
> To bank by computer, you will have to fill out a special application. When you set up the account, you will get a special password and a user identification number. This way, no one but you can look at your account or pay bills from your account.
>
> If you are interested in banking by computer, check with your bank. You will need to have a computer that has enough memory, and you will need access to the Internet. Talk to a bank officer to make sure that your computer system will meet their requirements.
>
> There may be a monthly service charge for banking by computer.

1. Why do you think banking by computer would be convenient? _____

   _____

   _____

2. Why is it so important for no one else to look at your bank account on the Internet? _____

   _____

   _____

3. Would it be worth $6 per month to you to be able to bank by computer? Why or why not? ___

   _____

   _____

   _____

4. What advantage do you think computer banking might have over banking by telephone

   (Activity 28)? _____

   _____

   _____

Name _____ Date _____

Electronic Banking Review

1. Tamika has $200.00 in her checking account. She writes checks for $45.00 and $30.00. She makes an ATM withdrawal of $50.00. What is her balance?

_____

2. Toh has $450.00 in his checking account. He pays the following bills by telephone: Jackson Water, $34.95; Energy Electric, $79.00; and Alabama Gas, $17.00. A monthly service charge of $3.95 is deducted for the bill-paying service. What is Toh's balance?

_____

3. Jamal has $319.00 in his checking account. He uses his debit card to make the following purchases: $250.00 to Chatham Apartments; $39.00 to Better Foods; and $26.00 to Rosario's Department Store. What is his balance?

_____

4. If the service charge on a bill-paying service is $3.95 a month, how many monthly bills would you need to pay by phone to save money on stamps? (Use the present cost of a stamp.)

_____

5. If the monthly service charge on a computer banking account is $6.95, what will be the total charge paid over a year?

_____

6. If you needed to use an ATM late at night, what things could you do to make sure the transaction is safe?

_____

_____

_____

7. If you were setting up a bank account today, which of the services included in this unit would you ask to have included? Which ones would you not use? Tell why you answered as you did.

_____

_____

_____

_____

_____

Name _____ Date _____

Credit Vocabulary

Credit	Buying something now and paying for it later.
Down payment	The part of total price of a purchase you pay right away. Also called a **deposit**.
Statement	A monthly report or bill from a credit-card company.
Balance due	The amount of money you owe the credit-card company.
Finance charge	Interest charged on the unpaid balance.
Minimum payment	The smallest payment you're allowed to make toward the balance due on your statement.

◆ **Fill in the Blanks.** Fill in each blank using a word or phrase from the box above. You will use each term only once.

1. Yetta couldn't pay her entire balance due, so she mailed the _____ to the credit-card company.

2. Kirk received his April _____ from the credit-card company yesterday.

3. Buying on _____ is a way to "buy now, pay later."

4. Elena bought a used car for $3,000. She made a _____ of $500 and took out a loan for the remaining $2,500.

5. 18 percent a year is the _____ used by many credit-card companies.

6. Matt paid $25 of the $45 he owed on his credit card. The _____ was then $20.

◆ **Matching.** Match each item in the left column with a word or phrase in the right column. Write the letter of the correct answer in each blank on the left. Use each choice only once.

7. _____ monthly report or bill   **(a)** credit

8. _____ least amount to be paid   **(b)** balance due

9. _____ a deposit paid on a purchase   **(c)** minimum payment

10. _____ interest charged   **(d)** statement

11. _____ buy now, pay later   **(e)** finance charge

12. _____ total amount owed   **(f)** down payment

Name _____ Date _____

What Is Credit?

Credit is a way to buy now with a promise to pay later. Some examples of credit are: student loans, home mortgages, car payments, and credit cards.

Credit can be good or bad, depending on how you use it.

- Credit helps you make large purchases you couldn't otherwise afford—like a car or a home. It can help in emergencies—for example, if you need car repairs and don't have the money.

- Credit is dangerous if you overspend. The payments mount up quickly, and interest is added if you don't pay the bill each month. You can soon get in over your head.

Before you pay for purchases with a credit card, stop and think. Remember, **credit is not more money.** You must still pay for the things you buy, and the payments will come due sooner than you think!

◆ **Directions:** Read each story. If it shows a good use of credit, write YES. If not, write NO. Then explain your answer.

1. Vangie had a job in a department store. The store gave her a store credit card. Vangie began shopping during her lunch hour, using the card for clothing, cosmetics, and other small items. At the end of the month, she was surprised to get a bill for $350.

   Did Vangie use credit wisely? _____ Why or why not? _____

   _____

2. Kahlil did not have a winter coat. He saw a good-looking, well-made coat marked down from $150 to $70. It was a good buy, but Kahlil did not have the money. So, he charged the coat and paid for it over 3 months. He ended up paying $78 for the coat, including the interest charges.

   Did Kahlil use credit wisely? _____ Why or why not? _____

   _____

3. Dean paid for all his gas and oil using a credit card. At the end of the month, he got a bill for $40.39. He mailed a check for $40.39 to the credit card company early enough so that he was charged no interest.

   Did Dean use credit wisely? _____ Why or why not? _____

   _____

Name _____ Date _____

How Credit Cards Work

Annual fees These vary a lot from card to card. Usually, cards that charge no annual fee charge higher interest rates. If you plan to pay off your balance each month, get a no-fee card. If you will carry a balance, get a low interest rate.

Credit limits When you get a credit card, you will be given a credit limit. You may not charge more than that amount. If you pay off your bills on time, you may get a higher limit.

The grace period On most cards, you have a period of time after making a charge before you are charged interest. No interest will be added to your balance during the grace period.

The minimum payment You will not have to pay the full amount you charge each month. You will be required to make a minimum payment. Then you will be charged interest on the balance.

Cash advances You may be able to get cash advances using your card. If you do, there will be a "transaction fee" as well as interest to pay on the amount.

1. Why do you think credit-card companies allow you to make a minimum payment instead of paying the full amount you owe? _____

   _____

   _____

2. Why do you think your credit limit will be low if you are a first-time card user? _____

   _____

   _____

3. Why do you think you should compare several credit cards before deciding on one to use?

   _____

   _____

4. You may get many credit-card applications in the mail. What would be a reason not to sign up for as many as you can? _____

   _____

5. Why is getting a cash advance not the best way to get the cash you need? _____

   _____

   _____

Name _____ Date _____

Getting a Credit Card

Let's say you are a young person just out of school. You are living on your own and working at your first job. You have decided that you would like to have a credit card. But, how do you go about getting one?

Mail offers You may get offers of credit cards in the mail. Some are good offers. Others are not. Read the information carefully. Study the features the card offers. Be sure to read the fine print carefully.

Telephone offers You may get calls about credit offers. Ask the caller to send you written information about the card. If the card is a good one, the company will be glad to do this. This way, you can look over the information carefully and make a good decision.

Your bank or credit union You will probably find a display of applications in the lobby. Ask a bank officer any questions you have about the card, its features, and how to apply for it.

Department stores or service stations Many stores and service stations offer credit cards. These may be good only at that company. These cards are usually easier to get than bank cards and carry a lower credit limit.

Offers of instant credit lines Beware of offers of instant money you may get in the mail. These probably charge high interest rates. Ignore ads that offer to get you a credit card "for a small fee." You should not need to pay to get a card.

1. What do you think you should do if you don't understand everything about a credit-card application you get in the mail? _____

   _____

2. Why do you think you should answer all questions on a credit-card application truthfully?

   _____

   _____

3. Why do you think it is best not to have more than one or two major credit cards? Why don't you need more? _____

   _____

4. Why do you think you should be careful where you keep your credit card? _____

   _____

   _____

Name _____ Date _____

Reading a Credit-Card Statement

> When you get a statement, check it over carefully. Make sure everything is correct before you make your payment.

◆ **Directions:** Read Tia Russem's credit-card statement below. Then answer the questions that follow it.

Charge Card Corp. Statement P.O. Box 193 Atlanta, GA 77402	Account No. 137 015 43009 Please pay by SEPT. 16 Amount you are paying _____

Account Summary

Last Month's Balance	Purchases This Month	Payments Received	Finance Charge	New Balance
80.75	345.09	80.75	0	345.09

To avoid finance charges on your next statement, your payment must equal the new balance and be received on or before the due date. The interest rate is $1\frac{1}{2}$% a month, or 18% a year.

Billing Date 8/25	Minimum Amount Due 17.00	Payment Due Date 9/16

Purchase Summary

Date	Item	Amount
8/3	Bryant's Department Store (clothing)	85.00
8/10	Direct Airlines	209.09
8/15	Elite Shoes	51.00

Make checks payable to Charge Card Corporation. Return the top part of this statement with your remittance. Address questions about this bill to Customer Service, Box 214, Atlanta, Georgia 77402.

1. What is Tia's account number? _____

2. How much did she charge this month? _____

3. Did she pay any finance charge this month? _____

4. What is Tia's minimum payment? _____

5. If Tia pays $50 this month, will she pay any interest next month? _____

6. If Tia mails her payment on September 23, will it get there on time? _____

7. To whom should Tia make out her check? _____

8. What two things should Tia send Charge Card Corporation by September 16?

    (1) _____

    (2) _____

9. If Tia has a question about her bill, where should she write? _____

Name _____ Date _____

Getting a Loan

Where to Get a Loan

If you need a loan, shop around. It's important to get that loan on the best possible terms. Places to get a loan include:

- Credit unions (They often have the lowest rates. You must belong to the credit union to do business there.)
- A bank (Check your own bank first. A bank often gives its customers a better rate. Then compare with other banks.)
- Finance companies (Avoid these. They usually charge high interest rates.)

How to Get the Loan

Once you've found the best deal you can, here's what to do:

1. Fill out a loan application.
2. The lender will check the information you give.
3. If the lender approves the loan, you will be asked to sign a loan agreement. Read it carefully first. Ask questions about any points you don't understand.
4. You will get the loan.
5. You must begin paying back the loan on time.

1. Name three places you would go in your area if you were shopping for a loan.

   _____

   _____

   _____

2. Why do you think some people get loans at high interest rates from places such as finance

   companies? _____

   _____

3. Why do you think it will cost you more to spread repaying a loan over a longer period of time?

   _____

   _____

4. Why do you think you should never sign a loan agreement before you read and understand

   everything in it? _____

   _____

Name _____ Date _____

Paying Off a Loan

> José owes the bank $1,066.20 for a $1,000 loan for a motorcycle. (Remember, the extra $66.20 is interest.) José must make 12 monthly payments of $88.85 to pay the loan back.
>
> José wants to keep track of how much he owes the credit union (his balance) each month. To do that, he must subtract his payment from the unpaid balance each month.

◆ **Directions:** Complete the payment chart below. The first two months are done for you.

AMOUNT JOSÉ BORROWED	$1,066.20
August payment	– $88.85
August balance	**$977.35**
September payment	– $88.85
September balance	**$888.50**
October payment	– $88.85
October balance	
November payment	– $88.85
November balance	
December payment	– $88.85
December balance	
January payment	– $88.85
January balance	
February payment	– $88.85
February balance	
March payment	– $88.85
March balance	
April payment	– $88.85
April balance	
May payment	– $88.85
May balance	
June payment	– $88.85
June balance	
July payment	– $88.85
July balance	$0
	The loan is paid off!

Name _____ Date _____

How to Save Money on Loans

There are three ways to get the best deal when shopping for a loan:
- Get the loan with the **lowest yearly finance charge**.
- **Make a large down payment** on the item. That way you can get a smaller loan and pay interest on a smaller amount.
- **Pay off the loan as soon as you can.** Since you pay interest on the unpaid balance each month, you'll pay more if you take more months to pay.

◆ **Directions:** Read each question below. Circle the letter of the credit deal that will cost less.

1. Jesse and Tranh each bought a home gym set for the same price. Who will pay less?
   (a) Jesse, who is paying over 6 months.
   (b) Sean, who is paying over 1 year.

2. Gayatri and Chris each bought identical new cars. Who will pay less?
   (a) Gayatri, with no down payment and an 18 percent finance charge.
   (b) Chris, with a $3,000 down payment and an 18 percent finance charge.

3. Tanisha bought a sound system. How will she pay less?
   (a) With store financing of 20 percent.
   (b) With a bank loan at 16 percent.

4. Derek needs to buy a new refrigerator. How will he pay less?
   (a) By paying cash.
   (b) By getting a credit union loan at 14 percent.

5. Sonya wants a new bicycle. How will she pay a lower finance charge?
   (a) By making a $50 down payment.
   (b) By making a $25 down payment.

6. Julio is shopping for a radio. He sees the same one at two stores. Which is the best deal?
   (a) Store #1: The radio costs $127 plus a finance charge of $14.
   (b) Store #2: The radio costs $119 plus a finance charge of $20.

7. Circle the letter of each action below that is a good way to get the best deal when shopping for credit.
   (a) Always make the smallest possible down payment.
   (b) Stretch payments over the longest time possible.
   (c) Don't shop around for credit. All rates are the same by law.
   (d) Make the largest down payment you can.
   (e) Pay off the loan as quickly as possible.
   (f) Buy from the store with the lowest finance rate if prices are about the same.

Name _____ Date _____

Making a Down Payment

> If you plan to pay for a large purchase over many months, most stores will ask for a **down payment** or **deposit**. A down payment is one large payment you make before you take the item home.
>
> The rest of the item's price is called the **balance**. The finance charge is figured on the balance. So your payments (and finance charges) will be less if you can make a larger down payment.

◆ **Directions:** Read the example in the box below. Then find the balance due for each item bought on credit in questions 1–3.

Example: Darnell bought an $800 home computer on credit. He made a down payment of $150.

      (a) What is the balance? ($800 − $150 = $650.)
      The balance is $650.

      (b) If the finance charge is $117, what will be the total cost he will pay?
      ($800 + $117 = $917.) He will pay $917 in all.

---

1. Jeanelle also bought a home computer on credit.
   The cash price was $800.
   She made a down payment of $50.

   (a) What is the balance? _____
   (b) If the finance charge is $135, what is the total cost she will pay? _____
   (c) Who will pay more for the same computer, Jeanelle or Darnell?
   _____

2. Carlo bought a new suit for work.
   The cash price was $175.99.
   He paid $25 as a down payment.

   (a) What is his balance? _____
   (b) If the finance charge is $27, what is the total cost he will pay for the suit? _____
   (c) If Carlo had paid a $50 down payment, would he have paid more or less in all for
   his suit? _____
   (d) If Carlo had paid cash, would he have ended up paying more, less, or the same
   for the suit? _____

3. Leslie bought a used boat for $3,000.
   She made a down payment of $375.

   (a) What is her balance? _____
   (b) If the finance charge is $540, what is the total cost she will pay for the boat?
   _____
   (c) How might she have paid less in finance charges? _____
   _____

Name _____ Date _____

Using Credit Responsibly

Often we hear of movie, rock, or sports stars with huge incomes who are filing for bankruptcy. No matter who you are, you can get into trouble using credit. It's tempting to buy more than you need with plastic. It's easy to ignore how hard the payments might be to make. And it's nice to get everything you want at once, instead of waiting until you can really afford it. Let's look at the purchases made by a young couple, the Moyas, using credit. See if you agree with their choices.

◆ **Directions:** Add the cash price and finance charge to find the credit price for each item the Moyas bought. Then find the total spent in each column. Finally, answer the questions below the chart.

Item	Cash Price	Finance Charge over 2 Years	Total Credit Price
Bedroom Set	$ 599.95	$ 135	
Living Room Set	$1,999.84	$ 485	
Dining Room Set	$ 485.99	$ 121	
Used Car	$4,999.95	$1,230	
TOTALS			

1. How much did the Moyas spend for all the items they bought on credit? _____

2. If they had saved their money and paid cash for the items, what would they have spent for all the items? _____

3. How much did the Moyas spend in total finance charges? _____

4. What advantages might there be to buying everything on credit as the Moyas did?

   _____

   _____

5. The Moyas' total monthly payments come to $421.00. How might a person get in trouble charging as many things as the Moyas did? _____

   _____

6. In what ways do you think the Moyas could have furnished their apartment for less money?

   _____

   _____

7. Do you think the Moyas used credit wisely? _____ Why or why not? _____

Name _____ Date _____

Credit Review

◆ **Directions:** Answer the following questions.

1. Name two ways in which credit can be helpful as a money-managing tool.

   _____

   _____

2. How might a person get into money trouble by misusing credit? _____

   _____

   _____

3. Keiko's credit card slips for March were for $15.85, $35.02, and $50.19. What were her total charges for March? _____

4. Carolyn bought a vacuum cleaner on credit. The cash price was $95. She made a down payment of $15. What was the balance? _____ With a finance charge of $15, what was the total cost of the vacuum cleaner? _____

◆ **Matching.** Match each item in the left column with a word or phrase in the right column. Write the letter of each correct term in the blanks on the left. You'll use each choice only once.

5. _____ monthly report or bill   **(a)** credit
6. _____ least amount to be paid   **(b)** balance due
7. _____ a deposit   **(c)** minimum payment
8. _____ interest charged   **(d)** statement
9. _____ buy now, pay later   **(e)** down payment
10. _____ amount of bill not yet paid   **(f)** finance charge

◆ **True/False.** Write **true** on the line in front of each true statement; write **false** in front of each false statement.

11. _____ You must always pay the total balance due on your statement each month.
12. _____ A credit card may be helpful if you have an emergency expense you can't pay for with cash.
13. _____ A credit card may save you money in the long run.
14. _____ The least expensive way to use a credit card is to make the minimum payment each month.
15. _____ You need not check your monthly statement since it is prepared by computer.
16. _____ All banks, credit unions, and finance companies must, by law, charge the same interest rates.
17. _____ An item paid for over a long period of time will probably cost you more than the same item paid for in cash.

Answer Key

Unit 1: Earning Your Money (The Paycheck)

Activity 1: Paycheck Vocabulary

1. net pay	4. gross pay	7. c	10. c
2. deductions	5. c	8. a	11. c
3. hourly pay	6. d	9. b	12. c

Activity 2: Finding Number of Hours Worked

1. 15   2. 6   3. 48   4. 13   5. 38   6. 14

Activity 3: Finding Gross Pay

1. $103.00	3. $59.50	5. $94.50	7. $77.55
2. $48.50	4. $150	6. $70.20	8. $119.00

Activity 4: Adding Deductions

Kim Chang's deductions: $99.29

LaToya Miller's deductions: $255.85

Sven Paulsen's deductions: $347.51

Activity 5: Finding Net Pay

1. Total deductions: $ 84.28	Net pay: $ 95.67
2. Total deductions: $261.48	Net pay: $327.71
3. Total deductions: $ 84.22	Net pay: $150.78
4. Total deductions: $354.18	Net pay: $430.91

Activity 6: Paycheck Review

1. 20	4. Gross pay: $312.00	5. c	8. a
2. $103.00	Total deductions: $47.80	6. d	9. e
3. $43.84	Net pay: $264.20	7. b	

Unit 2: Spending Your Money (A Checking Account)

Activity 7: Checking Account Vocabulary

1. balance	4. deposit slip	7. c	10. a
2. deposit	5. check	8. d	11. f
3. checkbook register	6. signature	9. e	12. b

Activity 8: Opening a Checking Account

In filling out the signature card, the term "title" means *Mr., Mrs., Ms.,* etc. Where it says "Type Name Below," have students print their names. Where it says "Sign Name Below," students should write their signatures.

1. The signature card is an official record of the account owner's signature against which the bank can compare the signature of anyone wishing to withdraw money. Thus, it prevents unauthorized persons from withdrawing money from an account.

2. In some cases, several persons may wish to have access to an account (for example, a husband and wife, a company account, etc.).

Activity 9: Writing Checks

Students fill in the sample check as directed.

Activity 10: Writing More Checks

Check each paper individually for correct form.

Activity 11: Making Deposits to a Checking Account

Check each deposit slip individually to see that it has been properly filled out. Total deposit should be $145.75.

Activity 12: Writing Checks and Deposit Slips I

Check each paper individually for correct form. Total deposit on #1 is $222.67.

Activity 13: Writing Checks and Deposit Slips II

Check each paper individually for correct form. Total deposit on #2 is $162.92.

Activity 14: A Checkbook Register

The balance after each check or deposit is:

A) $144.50	D) $552.63	G) $502.63
B) $603.50	E) $202.63	H) $437.63
C) $585.61	F) $43.63	I) $388.63

Activity 15: Checking Account Review

1. Total deposit: $46.06
2. Check for individual accuracy.
3. $34.51

4. $5.17
5. c
6. e

7. f
8. a
9. b

10. d

Unit 3: Saving Your Money (A Savings Account)

Activity 16: Savings Account Vocabulary

1. interest
2. account number
3. passbook

4. withdrawal
5. budget
6. deposit slip

7. f
8. d
9. c

10. e
11. a
12. b

Activity 17: Opening a Savings Account

1. *(any two)* It cannot be lost or stolen. You will earn interest. Your money is insured against loss.
2. Banks offer different types of accounts. You may be able to find an account that offers more interest. You should also look for an account with little or (preferably) no service charge. Banks may also differ in the services they offer.
3. A monthly service charge will reduce the actual amount of money you are making on the account.

Activity 18: Depositing Money in a Savings Account

Check each deposit slip individually to see that it has been properly filled out.

1. $76.25
2. $104.17
3. $48.67

Activity 19: Withdrawal Slips

Check each paper individually for proper completion.

Activity 20: A Savings Account Passbook

Balance at each date:

9/1	$50.00	11/3	$207.27
9/15	$100.00	11/15	$262.27
10/1	$150.00	12/1	$317.27
10/15	$205.00	12/15	$372.27
10/16	$190.00	12/17	$122.27
10/16	$192.27	12/29	$124.62
11/1	$247.27		

Activity 21: Planning Your Spending to Reach Your Goals

1. **Oscar's Budget:**
   Earnings: $232.90
   Lunches: $30
   Bus fare: $24
   Savings for car: $150
   Total: $204
   Money left for other spending: $28.90

2. **Yolanda's Budget:**
   Earnings: $319.00
   Lunches: $30
   Gas and oil: $50
   Savings for trip: $100
   Total: $180
   Money left for other spending: $139

Activity 22: Savings Build Up

1. After 4 weeks: $20
   After 26 weeks: $130
   After 52 weeks: $260
   After 5 years: $1,300
   Interest: $254

2. After 1 year: $200
   After 3 years: $600
   After 5 years: $1,000
   Interest: $193

3. After 6 months: $510
   After 1 year: $1,020
   After 5 years: $5,100
   Interest: $993

4. After 1 week: $30
   After 1 year: $1,560
   After 5 years: $7,800
   Interest: $1,518

Activity 23: Savings Account Review

1. Total deposited: $36.32
2. $600
3. 4 months
4. $390
5. $105
6. f
7. d
8. c
9. e
10. a
11. b

Unit 4: Automating Your Money (Electronic Banking)

Activity 24: Electronic Banking Vocabulary

1. debit card
2. service charge
3. PIN
4. ATM
5. transaction
6. c
7. e
8. d
9. a
10. b
11. b
12. a
13. c
14. e
15. d

Activity 25: Using an Automated Teller Machine (ATM)

1. To withdraw $75 from your checking account: Circle yes, no, no, checking; write in $75; circle no.

2. To deposit $200 to your savings account: Circle no, yes, no, savings; write in $200; circle no.

3. To check your balance in your checking account: Circle no, no, yes, checking, no.

Activity 26: ATM Safety

1. If your card is lost, no one can use it without the PIN. If you write the PIN on the card, anyone can find it and use it.

2. If someone makes you uncomfortable while you are using the ATM, cancel your transaction, remove your card, and leave. Make your transaction later, or find another ATM.

3. If you are using a drive-up ATM, keep your doors locked and your engine running. Roll down your window to make the transaction. If you are using a regular ATM, park and turn off your engine. Don't leave your keys in the car while you use the ATM.

4. These cards are another form of money. Anyone could use your debit card to make purchases in a store. He or she would need your PIN to use the card in an ATM. (That's why you should never write your PIN on your card or carry it in your wallet.)

Activity 27: Debit Cards

1. A debit card is easy to use. You don't have to write out the check or show ID.

2. You should record all uses of the debit card in your checkbook register as you would a check. Subtract each purchase amount from the balance in your account. If you don't do this, you will not know how much money is in your account.

3. The amount of each purchase made with a debit card is subtracted from your checking account right away. You will not receive a bill at the end of the month as you do with a credit card.

4. (Answers will vary.) If the user is careful to subtract each purchase from the checkbook balance, it should not be easier to overspend. If purchases are not recorded, the user might just keep on spending until the money is gone—risking overdrafts.

Activity 28: Banking by Telephone

1. When you write checks, some of them may not clear until days or weeks later. You might need to know the balance in your account that day. Or, you may need to know if a certain check or deposit had cleared. In general, checking your balance by phone is very convenient, since you can do it from any place at any time.

2. Figure out how many bills you normally pay by mail each month. Multiply the number of bills by the cost of a postage stamp to see how much you spend to pay bills by mail. In this example, it would cost you about $2.50 a month more to pay bills by phone ($30 a year).

3. You would not have to buy as many stamps, which could save trips to the post office. You would not have to write out the checks, address envelopes, and mail the bills.

4. You might want to know if someone you have written a check to has received and cashed the check. You may wish to put a "stop payment" on the check. (This would mean the person could not cash the check.) You might need to know the exact balance in your account.

Activity 29: Banking by Computer

1. You can bank by computer at any time of the day or night. You can check all your account balances, verify transactions, transfer funds, and pay bills quickly—without leaving home.

2. Your bank account is personal information. You would not want others to see your income and how you spend it. You also would not want anyone else trying to use your account illegally.

3. (Answers will vary.) The convenience would be worth it for many busy people. Also, those who are handicapped or who do not have ready access to transportation might find it useful. Students should remember that while $6 a month doesn't sound like much, the service is costing $72 a year.

4. You can see all your account information on the screen at once. This gives you a more complete overview of your transactions. It would also be helpful for the hearing-impaired.

Activity 30: Electronic Banking Review

1. $75.00

2. $315.10

3. $4.00

4. Divide $3.95 by the cost of a stamp. Whatever your answer is will be the "breakeven" number of bills you would have to pay by phone to make the service cost-effective. Any larger number of bills paid by phone would represent a savings.

5. $83.40

6. Some answers may include: Go to a well-lit ATM. Don't use the ATM if there are any people close by who make you uncomfortable. Take a friend with you. Cancel your transaction and leave if anyone makes you feel nervous.

7. Answers will vary.

Unit 5: Managing Your Money (Credit and Loans)

Activity 31: Credit Vocabulary

1. minimum payment
2. statement
3. credit
4. down payment
5. finance charge
6. balance due
7. d
8. c
9. f
10. e
11. a
12. b

Activity 32: What Is Credit?

1. No. She was buying a lot of small items that she probably didn't need, and she wasn't keeping track. She ended up with a large bill at the end of the month.

2. Yes. He got a good buy on a coat he needed. He saved money using credit because he was able to buy the coat on sale.

3. Yes. He had the convenience of buying on credit when he might not have had the cash to pay for the gas and oil. By paying off his balance quickly, he did not have an interest charge to pay.

Activity 33: How Credit Cards Work

1. Credit-card companies are pleased if you take a long time to pay off your balance. This way they can charge interest on the balance. This is how they make their money.

2. A first-time user has no track record of paying bills. The credit-card company does not know if you are reliable about paying your bills. They give you a low credit limit to start out, but may raise it if you prove yourself.

3. Credit cards vary a lot. You can save yourself a lot of money by shopping around and getting the best deal.

4. Having too many credit cards is one way people get into money trouble. If you make charges on a number of cards, you may quickly have too many bills to pay.

5. You will pay a transaction fee and also interest on what you get as an advance. If you do this several times a month, the charges will mount up.

Activity 34: Getting a Credit Card

1. There should be a toll-free number to call. Call and ask the service representative your questions. Be sure you understand everything about the features of the card.

2. The credit-card company will check all the information you give for accuracy before issuing you the card. If you give false information, you will not be issued the card.

3. Major credit cards can be accepted at a large number of places. If you are charging more than will go on two cards, you are probably charging too much.

4. Credit cards are money. If someone takes your card, he or she can use it to make charges.

Activity 35: Reading a Credit-Card Statement

1. 137 015 43009
2. $345.09
3. No
4. $17.00
5. Yes
6. No
7. Charge Card Corporation
8. 1) the top part of her statement   2) her payment (Explain the term *remittance* to students.)
9. Customer Service, Box 214, Atlanta, GA 77402

Activity 36: Getting a Loan

1. Answers will vary. Students should name local banks or credit unions.

2. They may have been turned down for a loan by a bank or credit union and have no other choice. Others may not realize they are paying more interest because they have not taken the time to shop around for their loan.

3. You pay interest on the unpaid balance. If you spread out the loan over a longer time, you will pay interest on a larger amount over a longer period of time.

4. You should be sure you understand the terms of the loan fully. Signing the paper may have consequences of which you are unaware.

Activity 37: Paying Off a Loan

Balances:

August	$977.35
Sept.	$888.50
Oct.	$799.65
Nov.	$710.80
Dec.	$621.95
Jan.	$533.10
Feb.	$444.25
March	$355.40
April	$266.55
May	$177.70
June	$88.85
July	$0

Activity 38: How to Save Money on Loans

1. a   3. b   5. a   7. Circle letters d, e, f
2. b   4. a   6. a

Activity 39: Making a Down Payment

1. (a) $750   (b) $935   (c) Jeanelle
2. (a) $150.99   (b) $202.99   (c) less   (d) less
3. (a) $2,625   (b) $3,540   (c) by making a larger down payment,
                                   by paying cash, or
                                   by paying off the boat more quickly

Activity 40: Using Credit Responsibly

The chart:

Total credit prices:	Bedroom set	$ 734.95
	Living room set	$2,484.84
	Dining room set	$ 606.99
	Used car	$6,229.95

Total cash price of all items:	$8,085.73
Total finance charge:	$1,971
Total credit prices:	$10,056.73

Questions:

1. $10,056.73
2. $8,085.73
3. $1,971
4. They have all the items they want to use and enjoy right away. If they make all the payments on time, they will be building a good credit rating.
5. If any unexpected emergencies arise (such as illness, loss of a job, pregnancy, car repairs, etc.), they may not be able to meet their payments. If that happened, the items might be repossessed and the Moyas' credit rating would be affected.
6. The Moyas might have bought used furniture. Good furniture can be found at garage sales or through want ads in the paper. They might have borrowed furniture. They might have bought only the essentials (e.g., a bed.) and bought new pieces gradually as they got the money. They might have waited for a sale, and added new pieces as they found good buys.
7. Answers will vary. If no problems arise, they may be able to meet all their financial obligations. If emergencies come up, they may have to charge more items. They will have a heavy debt load if this happens.

   Many people may not feel the Moyas' decisions were wise. It is not necessary to get everything at once. By buying the essentials and gradually acquiring other items, they could have avoided going into so much debt. They would then have extra money to use for savings or other needs.

Activity 41: Credit Review

1. Answers will vary. Credit can help you take advantage of sales. It can help meet emergencies for which you don't have the cash (e.g., a new hot water heater or an emergency airline ticket). You can avoid carrying a large amount of cash when shopping if you have a credit card. It may be easier to keep track of where you made purchases. It may be easier to return items that were charged than those paid for by check.
2. Answers will vary. It is easy to spend too much and become overextended with credit cards.

3. $101.06	7. e	11. false	15. false
4. $80, $110	8. f	12. true	16. false
5. d	9. a	13. true	17. true
6. c	10. b	14. false	

Share Your Bright Ideas with Us!

We want to hear from you! Your valuable comments and suggestions will help us meet your current and future classroom needs.

Your name_____Date_____

School name_____Phone_____

School address_____

Grade level taught_____Subject area(s) taught_____Average class size_____

Where did you purchase this publication?_____

Was your salesperson knowledgeable about this product?   Yes_____   No_____

What monies were used to purchase this product?

____School supplemental budget   ____Federal/state funding   ____Personal

Please "grade" this Walch publication according to the following criteria:

Quality of service you received when purchasing .... A   B   C   D   F
Ease of use... A   B   C   D   F
Quality of content... A   B   C   D   F
Page layout .... A   B   C   D   F
Organization of material .... A   B   C   D   F
Suitability for grade level .... A   B   C   D   F
Instructional value .... A   B   C   D   F

COMMENTS:_____

_____

What specific supplemental materials would help you meet your current—c

_____

Have you used other Walch publications? If so, which ones?_____

May we use your comments in upcoming communications?   ____Yes   ____No

Please **FAX** this completed form to **207-772-3105**, or mail it to:

Product Development, J.Weston Walch, Publisher, P.O. Box 658, Portland, ME 04104-0658

We will send you a **FREE GIFT** as our way of thanking you for your feedback. **THANK YOU!**